Questions and Answers

Levels I, II, and III

Neutron Radiographic Testing Method

Supplement to
Recommended Practice No. SNT-TC-1A
Book F

The American Society
for Nondestructive Testing, Inc.

first printing 1994
second printing with corrections 1999
third printing 08/10
fourth printing 10/14
fifth printing 08/19
sixth printing 09/21

Errata, if available for this printing, may be obtained from ASNT's website, asnt.org. Ebooks contain all corrections and updates, including the latest errata.

ISBN: 978-0-931403-14-9

Printed in the United States of America.

Published by:
The American Society for Nondestructive Testing Inc.
1711 Arlingate Lane
Columbus, OH 43228-0518
asnt.org

ASNT Mission Statement:
ASNT exists to create a safer world by advancing scientific, engineering, and technical knowledge in the field of nondestructive testing.

Code of Ethics:
The ASNT *Code of Ethics* was developed to provide members of the Society with broad ethical statements to guide their professional lives. In spirit and in word, each ASNT member is responsible for knowing and adhering to the values and standards set forth in the Society's Code. More information, as well as the complete version of the *Code of Ethics*, can be found on ASNT's website, asnt.org.

Table of Contents

References	**5**
Reference Usage List	**7**
Level I	**9**
Questions	**9**
Answers	**19**
Level II	**21**
Questions	**21**
Answers	**31**
Level III	**33**
Questions	**33**
Answers	**45**

Recommended Training References
Neutron Radiographic Testing Method

The following references were used in formulating the questions contained in this book.

A.* Berger, H. *Neutron Radiography*. Amsterdam, Netherlands: Elsevier Publishing Co. 1965.

B. Herz, R. *The Photographic Action of Ionizing Radiations*. New York: Wiley-Interscience. 1969.

C.* *Annual Book of ASTM Standards*. Volume 03.03, *Nondestructive Testing*. "Standard Method for Determining Image Quality in Direct Thermal Neutron Radiographic Examination," Standard ASTM E 545. Philadelphia, PA: American Society for Testing and Materials. Latest edition. (Reference numbers after the questions contain the standard number in them. Example: C-E545.)

D.* *Radiography in Modern Industry*. Rochester, NY: Eastman Kodak Co. 1969.

E. *Sensitometric Properties of X-ray Films*. Rochester, NY: Eastman Kodak Co.

F. Berger, H. "Neutron Radiography." *Annual Reviews of Nuclear Science, 21*. Palo Alto, CA: Annual Review Inc. 1971.

G. "Measurement of Neutron Flux and Spectra for Physical and Biological Applications." *NBS Handbook 72*. Washington DC: Superintendent of Documents. 1960.

H. Morgan, K. Z. and J. E. Turner, eds. *Principles of Radiation Protection*. New York: Wiley. 1967.

I. Wade, J. E. and G. E. Cunningham. *Radiation Monitoring — A Programmed Instruction Book*. USAEC —Division of Technical Information. 1967.

J. *Standards for Protection Against Radiation, Title 10, Rules and Regulations*. Part 20.

K. Glasstone. *Principles of Nuclear Reactor Engineering*. D. Van Nostrand Co., Inc. 1957.

L. *Chart of the Nuclides*. USAEC. 10th edition.

M. Hootman, H. E. *Estimation of ^{252}Cf Shielding Requirements*. D. P. 1232. Springfield, VA: US Department of Commerce. 1970.

N. "Kodak Film for Industrial Radiography." Supplement to *Radiography in Modern Industry*. Rochester, NY: Eastman Kodak Co. 1969.

O. Price, W. J. *Nuclear Radiation Detection*. New York: McGraw-Hill Book Co. 1958.

P. "Basic Radiation Protection Criteria." *NCRP Report No. 39*. Washington DC: NCRP Publications. January, 1971.

Q. "Safe Handling of Radioactive Material." *NBS Handbook 92*. Washington, DC: Superintendent of Documents.

R. Berger, H. *Practical Applications of Neutron Radiography and Gaging – STP 586*. Philadelphia, PA: American Society for Testing and Materials. 1973.

S.** Berger, H. "The Present State of Neutron Radiography and Its Potential." *Materials Evaluation*, Vol. 30, No. 3 (1972): 55-65.

Recommended Training References (continued)

T. Krishnamoorthy, P. N., et al. "Comparison of Various Neutron Personnel Dosimeters." *Neutron Monitoring for Radiation Protection Purposes.* Volume II, Proceedings of a Symposium, Vienna, Austria. 1973.

U. Richardson, H. D. *Industrial Radiography Manual, OE-84036.* Washington, DC: US Government Printing Office. 1968.

V.* McMaster, R.C., ed. *Nondestructive Testing Handbook,* first edition. Columbus, OH: The American Society for Nondestructive Testing, Inc. 1959.

W.* *Metals Handbook, Eighth Edition, Nondestructive Inspection and Quality Control, Volume 11.* Metals Park, OH: American Society for Metals. 1976.

X.** Beal, J. B. and R. L. Brown. "Advanced Radiographic Imaging Techniques." *Materials Evaluation*, Vol. 31, No. 7 (1973): 133-134.

Y. Fleischer, R. L., et al. *Nuclear Tracks in Solids.* University of California Press. 1975.

Z. Barton, J. P. and J. P. Perves. "Underwater Neutron Radiography with Conical Collimator." *British Journal of Nondestructive Testing*, Vol. 8, No.4 (1966).

AA.** Barton, J. P. "Divergent Beam Collimator for Neutron Radiography." *Materials Evaluation*, Vol. 25, No. 9 (1967): 45A-46A.

BB. Ilic, R., et al. "Microneutronography and Some Applications to Metallurgy." *Radiography with Neutrons.* Proceedings from a conference at the University of Birmingham, September, 1973.

CC. *Atomic Energy Review*, Vol. 15, No. 2 (1977). Murray Hill, NY: Unipub.

DD. Whittemore, W. L. "Fluxes, Beam Intensity, Collimation and Resolution for Neutron Radiography." General Atomic Company document, GA-9472, July 7, 1969.

* Available from The American Society for Nondestructive Testing, Inc.
**Available in photocopy form from the ASNT Information Center (800) 222-2768 or (614) 274-6003, X247.

Each question found in this book contains letter(s) and page number(s) in bold type and sometimes chapter or section number(s) immediately following the answers. For example,

17. Short wavelength electromagnetic radiation produced during the disintegration of nuclei of radioactive substances is called:

    a. X-radiation
    b. gamma radiation
    c. scatter radiation
    d. back-scattered radiation

    B.3

In this example, the letter "B" refers to Reference B in the list provided above. The "3" is the specific page in Reference B where the answer to the question can be found.

Reference Usage List
Neutron Radiographic Testing Method

Reference A: Total = 120
Level I (57)
Level II (43)
Level III (20)

Reference B: Total = 7
Level I (5)
Level II (0)
Level III (2)

Reference C: Total = 6
Level I (0)
Level II (0)
Level III (6)

Reference D: Total = 85
Level I (34)
Level II (34)
Level III (17)

Reference E: Total = 0
Level I (0)
Level II (0)
Level III (0)

Reference F: Total = 27
Level I (11)
Level II (9)
Level III (7)

Reference G: Total = 1
Level I (1)
Level II (0)
Level III (0)

Reference H: Total = 12
Level I (3)
Level II (9)
Level III (0)

Reference I: Total = 0
Level I (0)
Level II (0)
Level III (0)

Reference J: Total = 4
Level I (0)
Level II (1)
Level III (3)

Reference K: Total = 9
Level I (6)
Level II (0)
Level III (3)

Reference L: Total = 1
Level I (1)
Level II (0)
Level III (0)

Reference M: Total = 2
Level I (1)
Level II (1)
Level III (0)

Reference N: Total = 3
Level I (3)
Level II (0)
Level III (0)

Reference O: Total = 5
Level I (1)
Level II (2)
Level III (2)

Reference P: Total = 3
Level I (2)
Level II (1)
Level III (0)

Reference Usage List (continued)

Reference Q: Total = 3
Level I (1)
Level II (2)
Level III (0)

Reference R: Total = 1
Level I (0)
Level II (0)
Level III (1)

Reference S: Total = 15
Level I (0)
Level II (0)
Level III (15)

Reference T: Total = 1
Level I (0)
Level II (0)
Level III (1)

Reference U: Total = 1
Level I (0)
Level II (0)
Level III (1)

Reference V: Total = 3
Level I (0)
Level II (0)
Level III (3)

Reference W: Total = 2
Level I (0)
Level II (0)
Level III (2)

Reference X: Total = 1
Level I (0)
Level II (0)
Level III (1)

Reference Y: Total = 1
Level I (0)
Level II (0)
Level III (1)

Reference Z: Total = 1
Level I (0)
Level II (0)
Level III (1)

Reference AA: Total = 1
Level I (0)
Level II (0)
Level III (1)

Reference BB: Total = 1
Level I (0)
Level II (0)
Level III (1)

Reference CC: Total = 2
Level I (0)
Level II (0)
Level III (2)

Reference DD: Total = 1
Level I (0)
Level II (0)
Level III (1)

Level I Questions
Neutron Radiographic Testing Method

1. Neutron penetration is greatest in which of the following materials?

   a. hydrogenous material
   b. water
   c. lead
   d. boron carbide

   A.7

2. In general, by increasing the neutron energy from a neutron radiographic source:

   a. greater neutron penetration is achieved
   b. greater neutron radiographic contrast can be obtained
   c. radiographic exposure time can be reduced
   d. resolution can be increased

   A.10

3. The time required for one-half of the atoms in a particular sample of radioactive material to disintegrate is called:

   a. the inverse square law
   b. a curie
   c. a half-life
   d. the exposure time

   A.65, 136

4. Generally, the attenuation of neutrons by a given material is:

   a. reported to the Atomic Energy Commission
   b. greater for fast neutrons than thermal neutrons
   c. an indication of the quality of the X-radiographic technique
   d. appreciably greater for thermal and epithermal neutrons than for fast neutrons

   A.10

5. The mass absorption coefficients for thermal neutrons when plotted against regularly increasing atomic numbers of periodic elements presents a:

   a. blurred picture
   b. regularly increasing picture
   c. random picture
   d. dark picture

   A.6-7

6. Many of the absorption differences between neutrons and X-rays indicate clearly that the two techniques:

   a. cause radiation problems
   b. complement each other
   c. increase exposure speed
   d. fog radiographic film

   A.5-8

7. The neutron cross section is the term normally used to denote:

   a. the danger in handling radioactive material
   b. the absorbing power of a material for neutrons
   c. the atomic number of neutron reactor material
   d. radiation detection equipment

   A.10

8. The sharpness of the outline in the image of the radiograph is a measure of:

   a. subject contrast
   b. radiographic definition
   c. radiographic contrast
   d. film contrast

   D.67

9. The highest quality direct neutron radiographs obtainable today use:

   a. imaging screens using lithium-zinc sulfide as the imaging materials
   b. high-speed radiographic films
   c. dysprosium as an imaging screen
   d. gadolinium as an imaging screen

   A.59

10. When doing neutron radiography on radioactive materials, the materials are best handled:

    a. directly by personnel equipped with special protective clothing
    b. by remote handling equipment
    c. directly by personnel with special protective clothing except when radiographs are being made
    d. by the same methods used for nonradioactive materials

    A.137-138

11. Gadolinium conversion screens are usually mounted in rigid holders called:

    a. film racks
    b. cassettes
    c. emulsifiers
    d. diaphragms

A.90

12. The best high-intensity source of thermal neutrons is:

    a. a Cf-252 source
    b. an accelerator
    c. a nuclear reactor
    d. a Cf-252 source plus a multiplier

A.24-27; F.341

13. Scattered radiation caused by any material, such as a wall or floor, on the film side of the specimen is referred to as:

    a. primary scattering
    b. undercut
    c. reflected scattering
    d. back-scattered radiation

D.39-40

14. What has the highest thermal neutron absorption cross section?

    a. gold
    b. Indium
    c. gadolinium
    d. dysprosium

A.6-7

15. Conversion screens are used in neutron radiography:

    a. to convert neutron energy into ionizing radiation
    b. to increase the exposure time
    c. both a and b are reasons for using conversion screens
    d. neither a nor b is a reason for using conversion screens

A.46

16. A curie is the equivalent of:

    a. 0.001 mCi
    b. 1 000 mCi
    c. 1 000 MCi
    d. 100 MCi

K.61

17. Short wavelength electromagnetic radiation produced during the disintegration of nuclei of radioactive substances is called:

    a. X-radiation
    b. gamma radiation
    c. scatter radiation
    d. back-scattered radiation

B.3

18. A photographic record produced by the passage of neutrons through a specimen onto a film is called:

    a. a fluoroscopic image
    b. a radiograph
    c. an isotopic reproduction
    d. none of the above

A.46

19. Possible reactions that can occur when a fast neutron strikes a nucleus are:

    a. scattering and radiative capture
    b. microshrinkage and static charges caused by friction
    c. sudden temperature change and film contrast
    d. uniform thickness and filtered radiation

A.39

20. For inspection of radioactive objects or those that emit gamma radiation when bombarded with neutrons, a preferable detection method is the:

    a. direct exposure method
    b. transfer method
    c. isotopic reproduction method
    d. electrostatic-belt generator method

A.64, 86-87

21. Materials that are exposed to thermal neutron beams:

    a. must not be handled for at least 3 minutes after exposure has ceased
    b. must be stored in a lead-lined room
    c. may be radioactive after exposure to neutrons has ceased
    d. should be monitored by means of a neutron counter

A.91

22. Hydrogenous material has a:

    a. high macroscopic scattering cross section
    b. high absorption cross section
    c. high microscopic absorption cross section
    d. low microscopic scattering cross section

A.40

23. The penetrating ability of a thermal neutron beam is governed by:

    a. attenuation characteristics of the material being penetrated
    b. time
    c. source-to-film distance
    d. all of the above

A.9-10

24. A graph showing the relationship between film optical density and exposure is called:

    a. a bar chart
    b. a characteristic curve
    c. an exposure chart
    d. a logarithmic chart

D.53

25. The three main steps in processing a radiograph are:

    a. developing, frilling, and fixation
    b. developing, fixation, and washing
    c. exposure, developing, and fixation
    d. developing, reticulation, and fixation

D.80-81

26. Radiographic contrast in a neutron radiograph is least affected by:

    a. developer temperature
    b. radiographic exposure time
    c. radiographic beam collimation
    d. radiographic film fog

D.66

27. Higher resolution can be achieved in direct neutron radiography by:

    a. placing lead intensifying screen between a gadolinium screen and film
    b. increasing the L/D ratio of the collimation system
    c. increasing the exposure time
    d. increasing the distance between the object and the film cassette

F.340

28. The main reason for using neutron radiography in place of X-radiography is:

    a. lower cost
    b. higher resolution in all cases
    c. the ability to image objects and materials not possible with X-rays
    d. simpler radiographic procedure required than X-radiography

A.5

29. The best material for mounting specimens for neutron radiographic inspection is:

    a. cardboard
    b. plastic
    c. steel
    d. aluminum

A.90

30. Which of the following materials is best for making identification labels when using the neutron radiographic process?

    a. aluminum
    b. brass
    c. cadmium or gadolinium
    d. lead

A.6-7, 33

31. As a check on the adequacy of the neutron radiographic technique, it is customary to place a standard test piece on the source side of the cassette. This standard test piece is called:

    a. a reference plate
    b. a lead screen
    c. a penetrameter
    d. an image quality detector

A.95

32. A densitometer is:

    a. a meter used to measure neutron intensity
    b. an instrument used to measure film density
    c. a meter used to measure the density of a material
    d. a meter used to measure gamma content

B.91

33. The ability to detect a small discontinuity or flaw is called:

    a. radiographic contrast
    b. radiographic sensitivity
    c. radiographic density
    d. radiographic resolution

D.67

34. Movement, geometry, and screen contact are three factors that affect radiographic:

    a. contrast
    b. unsharpness
    c. reticulation
    d. density

D.21, 66

35. The difference between the densities of two areas of a radiographic film is called:

    a. radiographic contrast
    b. subject contrast
    c. film contrast
    d. definition

D.67

36. The selection of the proper type of film to be used for neutron radiographic examination of a particular part depends on the:

   a. thickness of the part
   b. material of the specimen
   c. neutron energy
   d. none of the above

D.75

37. When radiographing a part that contains a large crack, the crack will appear on the radiograph as:

   a. a dark, intermittent, or continuous line
   b. a light irregular line
   c. either a dark or light line
   d. a fogged area on the radiograph

B.459

38. Radiographic sensitivity, in the context of defining the minimum detectable flaw, depends on:

   a. the graininess of the film
   b. the unsharpness of the flaw image in the film
   c. the contrast of the flaw image on the film
   d. all of the above

D.19, 67, 69

39. An Image Quality Indicator is used to measure the:

   a. size of discontinuities in a part
   b. density of the film
   c. amount of film contrast
   d. quality of the radiographic technique

D.69

40. Unwanted inclusions in a part will appear on a radiograph as:

   a. a dark spot
   b. a light spot
   c. a generalized gray area of varying contrast
   d. either a dark or a light spot or area depending on the relative absorption ratio of the part material and the inclusion material

A.130

41. A sheet of cadmium with an opening cut in the shape of the part to be radiographed may be used to decrease the effect of scattered neutrons, which undercuts the specimens. Such a device is called a:

   a. mask
   b. filter
   c. back-scatter absorber
   d. lead-foil screen

A.60; D.40

42. The accidental movement of the specimen or film during exposure or the use of a source-film distance that is too small will:

   a. produce a radiograph with poor contrast
   b. make it impossible to detect large discontinuities
   c. result in unsharpness of the radiograph
   d. result in a fogged radiograph

D.20-21, 100

43. Dysprosium conversion screens emit:

   a. low-energy betas and gammas
   b. high-energy betas, low-energy gammas, and internal-conversion electrons
   c. beta particles only
   d. low-energy gamma rays only

L

44. Materials in common usage for moderation of fast neutron sources include:

   a. aluminum, magnesium, and tin
   b. water, plastic, paraffin, and graphite
   c. neon, argon, and xenon
   d. tungsten, cesium, antimony, and columbium

A.39-40

45. In the converter screen technique, the neutron image is produced by alpha, beta, or gamma radiation and it is thereby:

   a. used to measure neutron beam divergence
   b. externally cooled during the process
   c. photographically more detectable than the unconverted neutron image
   d. an important factor for determining Young's modulus of the material

A.46

46. Converter screen material characterized by lithium, boron, and gadolinium has little tendency to become radioactive but does:

   a. protect the radiographic film from excessive pressure
   b. recharge the focal point size of the neutron source
   c. filter and collimate the excess neutrons
   d. emit radiation immediately upon the absorption of a neutron

A.46-47

47. Gadolinium is frequently employed as a neutron absorber because of its:

   a. extremely low cost
   b. high neutron absorption for a given thickness
   c. ability to absorb gamma rays
   d. ability to diffract alpha particles

A.7

48. An excellent radiograph is obtained under given exposure conditions with a thermal neutron flux of 2×10^6 n/cm^2-s for 10 minutes. If other conditions are not changed, what exposure time would be required if the neutron flux was lowered to 1×10^6 n/cm^2-s?

a. 5 minutes
b. 10 minutes
c. 20 minutes
d. 30 minutes

A.102

49. Neutron converter screens should be inspected for flaws or dirt:

a. daily
b. each time they are used
c. occasionally
d. when flaws are detected on the radiograph

50. The primary advantage of using a Cf-252 source for neutron radiography is its:

a. portability
b. low cost per unit neutron flux compared to other neutron radiographic sources
c. high resolution
d. long useful life without source quality degradation

A.20-24; F.341-343

51. Neutron beams used in nondestructive testing normally contain:

a. alpha particles
b. positrons
c. gamma rays and neutrons
d. X-rays

F.242-246

52. In neutron radiography, L/D refers to the:

a. limiting neutron energy divided by the neutron density
b. largest neutron flux in the system divided by the beam diameter
c. distance from the neutron source to the object divided by the source diameter
d. distance from the neutron source to the picture divided by the beam diameter

F.339

53. A type of neutron beam collimator is a:

a. mean free path diaphragm
b. divergent beam collimator
c. polycellular field generator
d. neutron beam catcher

F.339

54. Commonly used converter screens are:

a. gadolinium, dysprosium, and indium
b. neodymium, plutonium, and technetium
c. gadolinium, lead, and indium
d. gold, silver, and cadmium

A.46-47, 67

55. In order to decrease geometric unsharpness:

a. neutrons should proceed from as small a source as other considerations will allow
b. neutrons should proceed from as large a source as other considerations will allow
c. the film should be as far as possible from the object being radiographed
d. the distance from the source to the material examined should be as small as practical

D.20-21

56. High-resolution gadolinium conversion screens are produced by:

a. flame spray techniques
b. being grown in large flat crystals
c. vacuum vapor deposition
d. large brazing systems

57. In order to increase the neutron beam intensity:

a. the L/D could be lowered
b. the neutron energy must be increased
c. the test specimen should be moved further from the film
d. a smaller source size could be used

F.338-339

58. Neutron exposure may be due to:

a. the direct beam from the neutron source
b. scatter radiation arising from objects in the direct beam
c. both a and b
d. neither a nor b

A.96

59. Gadolinium conversion screens emit:

a. gamma rays and conversion electrons
b. beta particles only
c. alpha particles and positrons
d. gamma rays only

A.60, 73-74

60. Which elements are commonly used in the indirect transfer method?

a. gadolinium and cadmium
b. rhodium and samarium
c. dysprosium and indium
d. cadmium and dysprosium

A.67

61. Which element is commonly used for direct neutron radiography?

    a. cadmium
    b. indium
    c. dysprosium
    d. gadolinium

A.49-64, 85

62. Neutron sensitive scintillators provide:

    a. high quality radiographs with long exposures
    b. low quality radiographs with long exposures
    c. low quality radiographs with short exposures
    d. none of the above

A.85

63. Lead is:

    a. a good neutron shield
    b. easily activated by neutrons
    c. a poor neutron absorber
    d. an efficient conversion screen

A.7

64. Neutron energy, exposure time, and film type are three important neutron radiographic parameters that can be controlled. What other parameter can be controlled?

    a. L/D
    b. conversion efficiency
    c. type of conversion screen
    d. both a and c

F.338-340, 350-351

65. The purpose of vacuum cassettes is to:

    a. eliminate scattered radiation
    b. block unwanted gamma radiation
    c. assure intimate film-to-foil contact
    d. protect parts from the radiation emitted by the conversion screen

D.38

66. In making a californium isotope exposure in an unshielded area, you find the dose rate 1.9 m (6 ft) from the source is 1 200 mR/h. What would be the dose rate at 7.3 m (24 ft)?

    a. 75 mR/h
    b. 100 mR/h
    c. 200 mR/h
    d. 300 mR/h

M.6

67. In developing film by hand technique without agitation:

    a. the radiograph will not show proper contrast
    b. it will be impossible to fix the radiograph
    c. there will be a greater fogging of the film
    d. there will be a tendency for each area of the film to affect the development of the area immediately below it

A.83

68. Film developed by automatic processors:

    a. will have improved detail of the image
    b. will have a general increase in the film density
    c. takes longer to develop than when processing manually
    d. will create less fog on the film

N.9-11

69. The emulsion side of a single-coated photographic X-ray-type film used for neutron radiography can be detected in the darkroom using standard safe lights as the:

    a. printed identifying marks on the emulsion side
    b. darker of the two sides
    c. duller and lighter color of the two sides
    d. printed identifying marks on the non-emulsion side

D.14

70. The purpose of film containers is to:

    a. protect the film from light
    b. protect the film from scratches
    c. neither a nor b
    d. both a and b

D.6

71. The two most common causes for excessively high density radiographs are:

    a. insufficient washing and overdeveloping
    b. contaminated fixer and insufficient washing
    c. overexposure and contaminated fixer
    d. overexposure and overdevelopment

D.147, 151

72. Single-emulsion high-resolution X-ray film is very good for neutron radiography because:

    a. it has a very thin emulsion
    b. it is sensitive to low-energy radiation and insensitive to high-energy radiation
    c. it is faster than other films
    d. both a and b

D.74; N.18-19

73. Which of the following materials is suited for construction of vessels or pails used to mix processing solutions?

a. stainless steel
b. aluminum
c. galvanized iron
d. tin

D.80

74. Excessive exposure of film to light prior to development of the film will most likely result in:

a. a foggy film
b. poor definition
c. streaks
d. a yellow stain

D.148

75. White crescent-shaped marks on an exposed X-ray film are most likely caused by:

a. crimping film after exposure
b. crimping film before exposure
c. sudden extreme temperature change while processing
d. a warm or exhausted fixer

D.148

76. Reticulation resulting in a puckered or netlike film surface is probably caused by:

a. crimping film after exposure
b. sudden extreme temperature change while processing
c. water or developer on unprocessed film
d. excessive object-film distance

D.150

77. Frilling or loosening of the emulsion from the base of the film is most likely caused by:

a. water or developer on unprocessed film
b. the low temperature of processing solutions
c. developer solution contamination
d. a warm or exhausted fixer solution

D.151

78. When the minute silver grains, on which the X-ray film image is formed, group together in relatively large masses, they produce a visual impression called:

a. air bells
b. graininess
c. reticulation
d. frilling

D.69

79. Static marks, which are black tree-like or circular marks on a radiograph, are often caused by:

a. the film being bent when inserted in a cassette or holder
b. foreign material or dirt embedded in screens
c. scratches on lead foil screens
d. improper film handling techniques

D.149

80. The purpose of agitating an X-ray film during development is to:

a. protect the film from excessive pressure
b. renew the developer at the surface of the film
c. disperse unexposed silver grains on the film surface
d. prevent reticulation

D.83

81. When manually processing films, the purpose of tapping the hangers sharply two or three times after the films have been lowered into the developer is to:

a. disperse unexposed silver grains on the film surface
b. prevent frilling
c. dislodge any air bubbles clinging to the emulsion
d. all of the above

D.83

82. The decrease in activity of the developer solution is compensated by:

a. constant agitation
b. maintaining processing solutions within the recommended temperature range
c. avoiding contamination from the wash bath
d. adding replenisher

D.84

83. The purpose of fixation is to:

a. remove all the undeveloped silver salts of the emulsion
b. leave the developed silver as a permanent image
c. harden the gelatin
d. all of the above

D.85-86

84. For the best results when manually processing film, solutions should be maintained within the temperature range of:

a. 65 °F and 75 °F
b. 65 °C and 75 °C
c. 75 °F and 85 °F
d. 75 °C and 85 °C

D.83, 85

85. Water spots on films can be minimized by:

a. the rapid drying of wet film
b. immersing wet film for 1 or 2 minutes in a wetting agent solution
c. using a fresh fixer solution
d. cascading water during the rinse cycle

D.87

86. The most suitable films for producing neutron radiography are:

a. red sensitive films
b. Polaroid™ films
c. industrial X-ray films
d. medical X-ray films

A.92

87. The normal development time for manually processing X-ray film is:

a. 12 to 18 minutes in processing solutions at 75 °F
b. 3 to 8 minutes in processing solutions at 75 °F
c. 12 to 18 minutes in processing solutions at 68 °C
d. 5 to 8 minutes in processing solutions at 68 °F

D.62

88. A properly exposed radiograph that is developed in a solution at a temperature of 58 °F will be:

a. overdeveloped
b. underdeveloped
c. fogged
d. damaged by frilling

D.82

89. An advantage of the fountain-pen type of ionization chamber used to monitor radiation received by personnel is that:

a. it provides a permanent record of accumulated dosage
b. it provides an immediate indication of dosage
c. it is the most sensitive detector available
d. all of the above

O.92

90. What radiation dose would be dangerous, if not fatal, if applied to the entire body in a short period of time?

a. 1.5 to 15 R (rem)
b. 25 to 70 R (rem)
c. 200 to 800 R (rem)
d. all of the above doses would most likely be fatal

H.460

91. The average thermal neutron flux that gives a dose of 100 mrem in 40 hours is:

a. 700 n/cm^2-s
b. 70 n/cm^2-s
c. 7 n/cm^2-s
d. 0.7 n/cm^2-s

A.137

92. When working with a neutron radiography facility, the radiation expected is:

a. gamma
b. beta
c. neutron
d. all of the above

A.81, 89

93. The intensity of neutron radiation is usually measured in:

a. roentgens
b. ergs
c. neutrons/cm-s
d. neutrons/cm

A.136

94. What does the term (R/h) refer to when speaking of intensity?

a. radiation limits for humans
b. roentgens per hour
c. X-rays per hour
d. radiation in hydrogen

A.136

95. Small amounts of exposure to neutrons or gamma rays:

a. may have a cumulative effect that must be considered when monitoring for maximum permissible dose
b. will be beneficial since they build up an immunity to radiation poisoning.
c. will have no effect on human beings
d. will have only a short-term effect on human tissues

K.535; P.69

96. Overexposure to neutron or gamma rays may cause damage to human:

a. blood and skin
b. skin
c. internal organs
d. all of the above

H.460-463

97. A general rule used to define an excessive amount of radiation exposure is:

a. anything above 0.4 R (rem) per week, although small amounts of radiation (0.4 R [rem] per week or less) are beneficial since they build up immunity to these rays
b. any dose over 5 R (rem) per week
c. any dose that causes a mid-range reading on a Geiger counter
d. any unnecessary exposure to radiation

P.63-64

98. A primary disadvantage of the fountain-pen type of ionization chamber used to measure the amount received by personnel is the:

a. delay necessary before the results of a measurement are known
b. inaccuracy of such devices in measuring scatter radiation
c. inability of such a device to provide a permanent record of exposure
d. cost of recharging such devices

Q.54

99. The exposure of personnel to X- and gamma radiation can be determined by means of:

a. film badges
b. dosimeters
c. radiation meters
d. all of the above

H.544-552

100. The intensity of gamma rays is measured in:

a. roentgens
b. ergs
c. roentgens per unit of time
d. H & D units

A.136

101. Divergent neutron beams:

a. do not obey the Inverse Square Law of distance
b. obey the Inverse Square Law of distance
c. use lead for shielding
d. none of the above

A.42

102. A radioactive source used for neutron radiography is:

a. PU-239
b. Co-60
c. Cs-137
d. Cf-252

F.343

103. The half-life of Cf-252 is:

a. 9 years
b. 2.6 years
c. 6 months
d. 47.5 years

F.342

104. Cf-252 is:

a. a spontaneous fission source
b. a fissile source
c. both a and b
d. neither a nor b

A.24

105. A normally desirable feature of a thermal neutron beam for neutron radiography is:

a. low gamma radiation intensity
b. relatively low, fast neutron intensity
c. low, angular divergence (so resolution capabilities for thicker objects can be good)
d. all of the above

A.81-82

106. To achieve uniformity in neutron radiographs, it is recommended that:

a. manual processing be used
b. automatic processing be used
c. it does not matter which process to use
d. the Shockly process be used

N.1

107. The approximate energy of a thermal neutron is:

a. 1 MeV
b. 0.026 eV
c. 12 KeV
d. 114 eV

A.9; K.15

108. The material that slows down neutrons is called:

a. a moderator
b. an accumulator
c. a limitor
d. none of the above

A.13; K.15

109. A good moderating material is:

a. water
b. iron
c. lead
d. all of the above

A.13; K.15

110. If the temperature decreases, the energy of a thermal neutron will:

a. increase
b. decrease
c. stay the same
d. none of the above

A.10, 110-111

111. A main disadvantage of gadolinium screens is:

a. that they are expensive
b. that they are magnetic
c. that they have poor sensitivity to neutrons
d. both a and b

112. The main advantage of a divergent beam collimator is:

a. that no dividing slats are used which could possibly cause lines on a radiograph
b. that there is a minimal neutron reflection if the sides of the collimator are made of a neutron absorbing material
c. that it is relatively simple to manufacture
d. all of the above

A.82-83

113. The main disadvantage of a divergent beam collimator is:

a. that less resolution is possible than with a parallel beam collimator of the same length
b. the large physical size of the collimator to achieve L/D ratios that are necessary for good resolution
c. that the small end of the collimator near the effective thermal neutron source is small, therefore minimizing the displacement of the moderator
d. all of the above

114. The definition of a neutron is:

a. the uncharged particle having a mass slightly greater than that of the proton
b. the uncharged particle having a mass slightly less than that of a proton
c. the positive charged particle having a mass slightly less than that of the proton
d. none of the above

B.13

115. To an Operational Health Physicist, the abbreviation, RBE means:

a. Rapid Biological Energy
b. Roentgen Background Embrittlement
c. Relative Biological Effectiveness
d. both a and b

G.2

116. The cross section is expressed in area units, the most common being the one that is equal to 10^{-24} cm^2. This is the:

a. femto
b. tero
c. barn
d. watt

A.10

117. The transfer exposure method is used because:

a. it is not sensitive to gamma radiation
b. it has greater radiographic sensitivity than the direct exposure method using gadolinium
c. it is faster than the direct exposure method
d. the screens used in this method emit only internal conversion electrons of about 70 keV

A.60, 64, 86-87, 89

118. Which of the following elements has the largest mass absorption coefficient for thermal neutrons?

a. boron
b. lead
c. gadolinium
d. copper

A.95

119. The term "macroscopic cross section (Σ)" is synonymous with the:

a. linear attenuation coefficient (μ) for neutrons
b. microscopic cross section (σ) for neutrons
c. mass attenuation coefficient (μ/ρ) for neutrons
d. cadmium ratio for neutrons

F.336; K.92

Level I Answers
Neutron Radiographic Testing Method

1.	**c**	31.	**d**	61.	**d**	91.	**a**
2.	**a**	32.	**b**	62.	**c**	92.	**d**
3.	**c**	33.	**b**	63.	**c**	93.	**c**
4.	**d**	34.	**b**	64.	**d**	94.	**b**
5.	**c**	35.	**a**	65.	**c**	95.	**a**
6.	**b**	36.	**d**	66.	**a**	96.	**d**
7.	**b**	37.	**a**	67.	**d**	97.	**d**
8.	**b**	38.	**d**	68.	**b**	98.	**c**
9.	**d**	39.	**d**	69.	**c**	99.	**d**
10.	**b**	40.	**d**	70.	**d**	100.	**c**
11.	**b**	41.	**a**	71.	**d**	101.	**b**
12.	**c**	42.	**c**	72.	**d**	102.	**d**
13.	**d**	43.	**b**	73.	**a**	103.	**b**
14.	**c**	44.	**b**	74.	**a**	104.	**a**
15.	**a**	45.	**c**	75.	**b**	105.	**d**
16.	**b**	46.	**d**	76.	**b**	106.	**b**
17.	**b**	47.	**b**	77.	**d**	107.	**b**
18.	**b**	48.	**c**	78.	**b**	108.	**a**
19.	**a**	49.	**b**	79.	**d**	109.	**a**
20.	**b**	50.	**a**	80.	**b**	110.	**b**
21.	**c**	51.	**c**	81.	**c**	111.	**a**
22.	**a**	52.	**c**	82.	**d**	112.	**d**
23.	**a**	53.	**b**	83.	**d**	113.	**b**
24.	**b**	54.	**a**	84.	**a**	114.	**a**
25.	**b**	55.	**a**	85.	**b**	115.	**c**
26.	**c**	56.	**c**	86.	**c**	116.	**c**
27.	**b**	57.	**a**	87.	**d**	117.	**a**
28.	**c**	58.	**c**	88.	**b**	118.	**a**
29.	**d**	59.	**a**	89.	**b**	119.	**a**
30.	**c**	60.	**c**	90.	**c**		

Level II Questions
Neutron Radiographic Testing Method

1. A nuclear reactor produces neutrons that have energies distributed from less than 0.01 eV to greater than 20 MeV. For most neutron radiography, the thermal energy group is used. This group has energies:

   a. below 0.01 eV
   b. 0 to 0.3 eV
   c. 0.3 to 10 000 eV
   d. 10 keV to 20 MeV
   e. >20 MeV

   A.10

2. The primary radiation mechanism for darkening X-ray film when the direct radiography process is used employing gadolinium screens is:

   a. alpha particles
   b. electrons
   c. gamma rays
   d. light emission

   A.60

3. Neutron radiography is an excellent tool for determining:

   a. the coating thickness of aluminum oxide on anodized aluminum
   b. voids in thick steel castings
   c. the integrity of thin plastic material within a lead housing
   d. voids in thick plastic sections

   A; F

4. Which of the following materials is the greatest attenuator of thermal neutrons?

   a. aluminum
   b. copper
   c. lithium
   d. iron

   A.5-7

5. Thermal neutrons that are relatively free from higher energy components are obtained by allowing neutrons from the source (reactor, radioactive source, accelerator) to pass through moderating material. This material may be:

   a. water
   b. paraffin
   c. graphite
   d. all of the above

   A.39ff

6. The sample part may become radioactive as the result of neutron activation during the neutron exposure. As a general rule, the level of radioactivity will be:

   a. high
   b. very low
   c. dangerously high
   d. none of the above

   A.91

7. The absorption of neutrons in an object depends upon:

   a. the neutron cross section
   b. the nuclides in the object
   c. the object thickness
   d. all of the above

   A.5

8. The energy of the neutrons in a neutron beam:

   a. is the same for all neutron beams
   b. does not affect the radiographic parameters
   c. has no bearing on neutron absorption in the material to be radiographed
   d. directly influences the choice of usable conversion screens

9. Gadolinium and cadmium have:

   a. a flat neutron cross section across all energies
   b. a cross section which peaks at thermal energies
   c. a high thermal neutron cross section which drops off rapidly at higher energies
   d. their highest neutron cross section at 1.4 eV

   A.11

10. Which of the following conversion screens has the longest half-life?

    a. dysprosium
    b. indium
    c. cadmium
    d. gadolinium

A.47

11. The unique and important properties of the neutron in neutron radiography come primarily from the fact that it is a nuclear particle that is electrically neutral. The lack of electric charge means that its electrostatic interaction with the atom's electrons is:

    a. almost complete
    b. scattered
    c. almost completely negligible
    d. such that the electrons and neutrons combine with each other

D.10

12. The rare earths are frequently used in neutron radiography. An interesting feature of this family of elements is:

    a. that they have nearly identical chemical properties and are, therefore, difficult to tell apart
    b. that they make up about one-sixth of all naturally occurring elements, but the entire group occupies only one position in the Periodic Table
    c. that they have large absorption cross sections for neutrons
    d. all of the above

A.90

13. A thin sheet of gadolinium foil, in intimate contact with photographic film during neutron exposure, will increase film density because:

    a. it fluoresces and emits visible light, which helps expose the film
    b. it emits electrons that darken the film
    c. it absorbs the scattered radiation
    d. it prevents back-scattered radiation from fogging the film

A.60

14. For direct neutron radiography, precaution has to be taken to reduce the gamma ray background of most sources as this tends to obscure the neutron radiograph. This may be done by:

    a. using filtration of gamma rays
    b. increasing the source-to-film distance
    c. both a and b
    d. neither a nor b

A.30, 82

15. The transfer exposure technique, sometimes called the indirect technique, uses neutron converters that have a reasonably long radioactive half-life. Which of the following would be a good candidate?

    a. indium
    b. dysprosium
    c. both a and b
    d. neither a nor b

A.47

16. Which of the following neutron radiography converter foils cannot be used for transfer or indirect radiography?

    a. dysprosium
    b. indium
    c. gadolinium
    d. gold

A.47

17. The quality of the results from a neutron radiographic facility is best determined by:

    a. reference standards
    b. image-quality indicators
    c. neutron flux measurement
    d. densitometer readings

A.95; D.69-73

18. Flaws in the imaging screens can be separated from actual flaws in a part being radiographed by:

    a. performing a photographic copy of the original neutron radiograph using X-ray duplicating film
    b. comparing a neutron radiograph of the parts to a blank neutron radiograph of the same imaging screen with no parts in place
    c. increasing the exposure time of the radiograph
    d. decreasing the temperature of the developer solution

D.33, 148

19. The best method for determining how resolution affects the ability to interpret the radiograph of parts is by:

    a. radiographing parts that have known measured defects
    b. using a penetrameter
    c. using a penetrameter containing cadmium wedges
    d. using a penetrameter containing plastic wedges

20. The slope of a straight line joining two points of specified density on the characteristic curve of a film is known as the:

a. speed of the curve
b. latitude
c. average gradient
d. density

D.126

21. The range of densities which are satisfactory for interpretation is a measure of the:

a. subject contrast of a radiograph
b. sensitivity of a radiograph
c. latitude of a radiograph
d. definition of a radiograph

D.124-127

22. The transmission of neutrons by a material varies:

a. directly with the square of the distance from the source
b. directly with the thickness of the material
c. inversely with the amount of scattering in the material
d. exponentially with the thickness of the material

A.109

23. Which of the following is not a factor in determining subject contrast?

a. the nature of the specimen
b. the neutron energy
c. the type of film used
d. the intensity and distribution of the scattered radiation

D.66-67

24. When viewing a radiograph, an image of the back of the cassette superimposed on the image of the specimen is noted. This is most likely due to:

a. undercut
b. overexposure
c. neutron intensity being too high
d. back scatter

D.39-40

25. A qualitative term often used to indicate the size of the smallest detail which can be seen in a radiograph is:

a. radiographic sensitivity
b. radiographic definition
c. radiographic contrast
d. subject contrast

D.67

26. The quantitative measure of film blackening is referred to as:

a. definition
b. photographic density
c. film contrast
d. radiographic contrast

D.51

27. A curve that relates density to the logarithm of exposure or of relative exposure is called:

a. a sensitivity curve
b. a density-exposure curve
c. a characteristic curve
d. an X-ray intensity curve

D.53

28. Subject contrast is affected by:

a. thickness differences in the specimen
b. neutron energy
c. scattered radiation
d. all of the above

D.66

29. Contrast enhancement of a neutron radiograph can be achieved by:

a. using photographic techniques with X-ray duplicating film
b. increasing the radiography system resolution
c. varying the object-to-film distance
d. increasing the gamma radiation reaching the film from the source

D.112-113, 125

30. The uniformity of a neutron radiograph is best determined by measuring the density of a:

a. neutron radiograph at several locations with many parts in place
b. neutron radiograph at several locations with no parts in place
c. photographic copy of a neutron radiograph with many parts in place
d. photographic copy of a neutron radiograph with no parts in place

31. The density measurement in a neutron radiograph that is of most interest in determining part flaws is the density measurement:

a. at the center of the film
b. at the corner of the film
c. at the edge of the film
d. in the area of interest within the parts being radiographed

32. A neutron radiograph beam from a reactor was measured to have a thermal neutron flux of 3.4×10^6 n • cm^{-2} • s^{-1} and a gamma rate of 26 R/h. Which of the following would be the neutron-to-gamma ratio in units of n • cm^{-2} • mR^{-1}?

    a. 4.7×10^2
    b. 4.7×10^5
    c. 2.1×10^5
    d. 2.1×10^2

    A.81

33. In direct neutron radiography using gadolinium, cleanliness is essential when handling film cassettes because:

    a. dust or lint between the film and the gadolinium shows as a flaw in the radiograph
    b. dust on the outside of the film cassette is always visible in the neutron radiograph
    c. dust is highly absorbent to neutrons
    d. dust greatly affects the development of X-ray film

    D.148

34. A dysprosium conversion screen is exposed in a thermal neutron beam. After decay time of 6.9 hours:

    a. 3/4 of the original activation will remain
    b. 1/2 of the original activation will remain
    c. 1/8 of the original activation will remain
    d. there will be no change from original activation levels

    A.47, 65

35. When sharp, black, bird-foot shaped marks appear on the film in areas where there are no possible discontinuities, they are probably caused by:

    a. prolonged development in an old developer
    b. exposure of the film by natural cosmic radiation
    c. static charges caused by friction
    d. inadequate rinsing after fixing

    D.148

36. Images of discontinuities close to the source side of the specimen become less clearly defined as the:

    a. source-to-object distance increases
    b. thickness of the specimen increases
    c. size of the neutron source decreases
    d. thickness of the specimen decreases

    D.18-21

37. The amount of unsharpness of a radiograph is:

    a. directly proportional to the object-to-film distance and inversely proportional to the size of the source
    b. directly proportional to the size of the source and inversely proportional to the source-to-object distance
    c. inversely proportional to the object-to-film distance and directly proportional to the source-to-object distance
    d. inversely proportional to the size of the source and the object-to-film distance

    A.18-21

38. The most commonly used converter material is:

    a. copper
    b. tungsten
    c. gold
    d. gadolinium

    A.60, 84

39. As the effective energy of the radiation from the conversion screen increases:

    a. film graininess increases
    b. film graininess decreases
    c. radiographic definition decreases
    d. film speed decreases

    A.60, 85

40. A general rule governing the application of the geometric principles of shadow formation states that the:

    a. neutrons should proceed from as large a source area as other considerations will allow
    b. distance between the source and material should be as small as practical
    c. film should be as far as possible from the object being radiographed
    d. axis of the beam should be as nearly perpendicular to the film as possible, to preserve spatial relationships

    D.18-21

41. A neutron radiograph made with an exposure of 12 minutes has a density of 0.8 in the region of maximum interest. It is desired to increase the density in this area to 2.0. By reference to a characteristic curve of the film, it is found that the difference in Log E between a density of 0.8 and 2.0 is 0.76. The antilogarithm of 0.76 is 5.8. What must be the new exposure time to produce a radiograph of density 2.0?

    a. 30 minutes
    b. 21.12 minutes
    c. 69.6 minutes
    d. 16 minutes

    D.55

42. Which of the following factors will not materially influence the image density of a neutron radiograph?

    a. the type of film used
    b. the size of the film
    c. the type of conversion screen used
    d. the exposure time

A; D

43. X-ray films with large grain size:

    a. will produce radiographs with better definition than film with small grain size
    b. have slower speeds than those with a relatively small grain size
    c. have higher speeds than those with a relatively small grain size
    d. will take longer to expose properly than film with a relatively small grain size

D.133

44. The uneven distribution of developed grains within the emulsion of a processed X-ray film causes the subjective impression of:

    a. graininess
    b. streaks
    c. spots
    d. white scum

D.132

45. An X-ray type film used for neutron radiography having wide latitude also has, by definition:

    a. poor definition
    b. low contrast
    c. high speed
    d. none of the above

D.126-127

46. The gradient of a characteristic curve for a photographic film for neutron radiography is the greatest as a density of:

    a. <0.5
    b. 0.5
    c. 1.0
    d. >1.0

D.126-127

47. For practical purposes, the shape of the characteristic curve of an X-ray film is:

    a. independent of the type of film used
    b. independent of the energy of the neutron beam
    c. drastically changed when neutron energy is changed
    d. primarily determined by the subject contrast

D.126-127

48. The interval between the time a film is placed in a fixer solution and the time when the original diffuse, yellow milkiness disappears is known as the:

    a. clearing time
    b. fixing time
    c. hardening time
    d. oxidation time

D.86

49. Improper geometric factors, poor contact between film and conversion screen, and graininess of film are possible causes of:

    a. high film density
    b. poor definition
    c. fogged film
    d. increased contrast

D.147-149

50. It is known that the density of a film increases with increasing exposure up to a maximum value. Increasing the exposure beyond this point results in an actual decrease of density. This phenomena is referred to as:

    a. density-intensity turnabout
    b. subject contrast inversion
    c. film contrast inversion
    d. reversal

B.136

51. The activity of the fixer diminishes after being used for a period of time because:

    a. the active ingredients evaporate
    b. the active ingredients are absorbed by the radiograph
    c. the fixer solution accumulates soluble silver salts
    d. the active ingredients settle to the bottom of the tank

D.86

52. Developer solution should be discarded when the quantity of replenisher added equals:

    a. the original quantity of developer
    b. 2 to 3 times the original quantity of developer
    c. 5 to 6 times the original quantity of developer
    d. 10 times the original quantity of developer

D.85

53. Agitation of the X-ray film during the development process by means of mechanical stirrers or circulating pumps may:

    a. speed the developing cycle
    b. help replenish the developer
    c. cause undesirable preferential flow of developer along certain paths
    d. cause reticulation

D.84

54. In processing radiographs, the hourly flow of water in the wash tank should be:

    a. 2 to 3 times the volume of the tank
    b. 4 to 8 times the volume of the tank
    c. at least 151 L (40 gal) per hour
    d. varied continuously in proportion to the number of radiographs being developed

D.86

55. The slope (steepness) of a characteristic curve is a measure of the:

    a. subject contrast
    b. radiographic definition
    c. radiographic contrast
    d. film contrast

D.53-30

56. As the development time increases:

    a. the characteristic curve grows steeper and moves to the left
    b. the characteristic curve grows steeper and moves to the right
    c. the characteristic curve remains the same in shape but moves to the left
    d. there is little effect on the characteristic curve

D.53-60

57. Which of the following instruments would most likely by used to detect small leaks in a radiation barrier?

    a. a film badge
    b. a fountain pen type of ionization chamber
    c. a Geiger counter
    d. a dosimeter

A.115, 138; Q.54-55

58. The quantity of neutron radiation upon an area of film for an exposure:

    a. is the product of the neutron flux and time
    b. is the neutron flux
    c. varies exponentially with time
    d. varies inversely proportional with time

A.65

59. The energy of the neutron is expressed by which of the following units of measurement?

    a. curie
    b. roentgen
    c. half-life
    d. electron volts

A.10, 135

60. Neutrons for fast-neutron radiography are obtainable from:

    a. accelerators
    b. radioactive sources
    c. reactors
    d. all of the above

A.9-38

61. "Macroscopic cross section" as applied to neutrons is analogous to for γ rays:

    a. cross section
    b. mean free path
    c. attenuation coefficient
    d. wavelength

A.7, Table 1

62. For high quality radiographs, a 25-micrometer thick gadolinium screen combined with fine-grain X-ray film requires an exposure of about:

    a. 3×10^3 n/cm^2
    b. 3×10^5 n/cm^2
    c. 3×10^7 n/cm^2
    d. any of the above

F.349

63. Neutron monitoring outside a radiography exposure area may be done with:

    a. a Geiger counter
    b. a "cutie-pie" (ionization chamber)
    c. a BF_3 proportional counter
    d. all of the above

H.544-549, 557

64. Special neutron-sensitive film dosimeters:

    a. should be worn by neutron radiographers
    b. employ a film-screen combination principal similar to radiography
    c. also give a measure of γ dose
    d. all of the above

H.220

65. For the same beam intensity, which of the following will give the largest biological dose?

    a. fast neutrons
    b. thermal neutrons
    c. $\gamma\alpha\mu\mu\alpha$ rays
    d. all are equal by definition

A.137

66. For fast-neutron monitoring, a thermal-neutron-sensitive radiation counter may be used in conjunction with:

a. a moderator
b. an ionization chamber
c. a conversion foil
d. a collimator

H.558-559

67. If the biological dose rate at 1.5 m (5 ft) from a point source of radiation is 10 mrem/h, then the minimum permissible distance at which a worker may remain for continuous (full work week) exposure is about: (Assume maximum exposure of 100 mrem/week)

a. 3 m (10 ft)
b. 6 m (20 ft)
c. 15.2 m (50 ft)
d. 30.5 m (100 ft)

M.6

68. In a radiation area where the dose rate is established to be 25 mrem/h, a worker during a work week may spend no more than: (assume minimum exposure of 100 mrem/week)

a. 6 minutes
b. 1 hour
c. 4 hours
d. 8 hours

J

69. Upon exiting from a radiation zone, you begin a G-M survey of an object. The needle begins to rise and then suddenly drops to zero. You should assume that the:

a. G-M tube has ruptured
b. G-M jammed because of too much radiation
c. G-M probe has become grossly contaminated
d. rubbers are not contaminated

O; P; Q

70. The G-M cannot be used to establish dose rates because:

a. the G-M is not reliable and tends to drift
b. the beta shield is too thick on the G-M
c. the G-M does not account for the degree of ionization or energy absorption of ionizing particles or photons
d. the G-M is designed to detect only alpha contamination on scintillators

H.548

71. A direct reading pocket dosimeter may be sensitive to what radiation?

a. β
b. γ
c. γ, n
d. either a or b

H.546; O.89-93

72. A film dosimeter useful for neutron radiographers will be sensitive to what radiation?

a. α, β, γ, n
b. α, β, γ
c. β, γ
d. β, γ, n

H.551

73. Thermal-neutron-sensitive radiation counters usually contain:

a. gadolinium
b. cadmium
c. boron
d. lead

H.559

74. If 2 mm (0.08 in.) of plastic attenuates a thermal neutron beam by a factor of 2, then 2 mm (0.8 in.) will attenuate it by approximately a factor of:

a. 10
b. 100
c. 1 000
d. 10 000

A.109

75. The high attenuation of thermal neutrons by plastics is due primarily to:

a. scattering
b. absorption
c. both a and b
d. none of the above

A.6

76. Photographic density is a quantitative measure of:

a. film thickness
b. film weight
c. film blackening
d. object opacity

D.51

77. Film exposed to a density of 2 will transmit what percentage of the light incident upon it?

a. 50%
b. 25%
c. 2%
d. 1%

D.51

78. An image-quality indicator is:

a. a scanning densitometer
b. a chemical stain
c. a penetrameter
d. all of the above

D.69

79. Radiography using the transfer method implies that the imaging screen:

a. is placed behind the film
b. is placed in front of the film
c. is very thin
d. becomes radioactive

A.64ff

80. In the list below, the best filter material for making an epithermal neutron beam is:

a. cadmium
b. water
c. boron
d. lead

A.11

81. Fast neutron attenuation:

a. increases significantly with increasing atomic mass
b. decreases significantly with increasing atomic mass
c. has significant random variations with atomic mass
d. is similar for most materials

A.10-11

82. Which of the following is most widely used in the detector for imaging fast neutrons?

a. hydrogenous material
b. gold
c. aluminum
d. lithium

F.358

83. The linear attenuation coefficient for neutrons is described in the following equation: $\mu = N\sigma_t$. Which of the following statements is true?

a. N = number of nuclei per cm^3 of attenuating material
b. σ_t = total cross section (cm^2), equal to the sum of absorption and scattering cross sections ($\sigma_a + \sigma_s$)
c. σ_a = absorption cross section
d. all the above

F.336

84. Real-time imaging of thermal neutron radiography can be performed with which of the following detectors?

a. gadolinium
b. dysprosium
c. zinc sulfide + lithium fluoride
d. europium

F.352

85. A neutron beam undergoes which of the following interactions when penetrating matter?

a. scatter
b. pair production
c. Thompson effect
d. compton scattering

F.336

86. A Geiger-Muller instrument is a:

a. pocket sized dosimeter
b. scintillation counter
c. hand-held survey meter
d. proportional counter

H.546

87. The thermalization factor is the:

a. ratio between the total 4π fast neutron yield and the peak thermal neutron flux
b. dose of ionizing radiation that can be absorbed per unit volume
c. mean square distance while slowing down
d. mean time spent diffusing in a test object before being captured in a detector

F.341

88. Energy classification places epithermal neutrons in the range:

a. below 0.01 eV
b. from 0.3 eV to 100 keV
c. from 0.3 to 10 000 eV
d. from 10 keV to 20 MeV

A.10

89. A shutter for turning the neutron beam ON and OFF at a nuclear reactor can be accomplished by:

a. the electronic circuitry
b. a small mechanical camera shutter
c. boral shutters
d. aluminum shutters

A.84

90. Gadolinium screens have been shown to resolve high-contrast images separated by distances as small as:

   a. 2.5 mm (0.1 in.)
   b. 1 mm (0.04 in.)
   c. 0.1 mm (0.004 in.)
   d. 0.01 mm (0.0004 in.)

A.84

91. The best radiographic resolution and contrast capability for direct exposure radiography has been obtained with:

   a. gadolinium screens
   b. indium
   c. silver
   d. cadmium

A.84-85

92. Radiography requiring utmost speed and no more than 0.05 mm (0.002 in.) resolution would require which of the following detectors?

   a. gadolinium screens
   b. scintillator screens
   c. dysprosium screens
   d. silver screens

A.85

93. In which of the following neutron beams does the intensity follow the inverse square law?

   a. divergent beam
   b. soller slit beam
   c. neutron spectrometer beam
   d. monochromatizing crystal beam

A.42

94. Neutron radiography:

   a. complements X-radiography
   b. can discriminate between neighboring elements such as boron and carbon
   c. can inspect large thicknesses of heavy metals
   d. all of the above

A.5

95. Neutron radiography extends radiographic capability for detecting cracks in small:

   a. aluminum pins
   b. iron pins
   c. magnesium pins
   d. plutonium pins

A.120

96. Film fogging during radiography of radioactive materials (5 000 R/h at 31 cm [1 ft]) is minimized by using:

   a. direct X-radiographic methods
   b. pinhole autoradiography
   c. transfer method with dysprosium screens
   d. photography

A.124-126

97. What radiography technique should be used for obtaining improved penetration of 20% Pu, U-235 enriched fuel pellets, ~ 6 mm diameter?

   a. thermal neutron with gadolinium screen
   b. epithermal neutron and indium screen
   c. epithermal neutron and dysprosium
   d. autoradiography

F.356-357

98. Neutron radiography can be used for inspecting which of the following applications?

   a. presence of explosive in a metal device
   b. presence of foreign material such as oil
   c. lubricants in metal systems
   d. hydrogen content in metals
   e. all of the above

F.359

99. Disadvantages of the transfer technique include that it:

   a. is time consuming
   b. requires many imaging foils
   c. provides lesser resolution than the gadolinium direct method
   d. costs more to use dysprosium foils than gadolinium foils
   e. all of the above

A.85

Level II Answers
Neutron Radiographic Testing Method

1.	**b**	26.	**b**	51.	**c**	76.	**c**
2.	**b**	27.	**c**	52.	**b**	77.	**d**
3.	**c**	28.	**d**	53.	**c**	78.	**c**
4.	**c**	29.	**a**	54.	**b**	79.	**d**
5.	**d**	30.	**b**	55.	**d**	80.	**a**
6.	**b**	31.	**d**	56.	**a**	81.	**d**
7.	**d**	32.	**b**	57.	**c**	82.	**a**
8.	**d**	33.	**a**	58.	**a**	83.	**d**
9.	**c**	34.	**c**	59.	**d**	84.	**c**
10.	**a**	35.	**c**	60.	**d**	85.	**a**
11.	**c**	36.	**b**	61.	**c**	86.	**c**
12.	**d**	37.	**b**	62.	**d**	87.	**a**
13.	**b**	38.	**d**	63.	**c**	88.	**c**
14.	**a**	39.	**c**	64.	**d**	89.	**c**
15.	**c**	40.	**d**	65.	**a**	90.	**d**
16.	**c**	41.	**c**	66.	**a**	91.	**a**
17.	**b**	42.	**b**	67.	**a**	92.	**b**
18.	**b**	43.	**c**	68.	**c**	93.	**a**
19.	**a**	44.	**a**	69.	**b**	94.	**d**
20.	**c**	45.	**b**	70.	**c**	95.	**d**
21.	**c**	46.	**d**	71.	**d**	96.	**c**
22.	**d**	47.	**b**	72.	**d**	97.	**b**
23.	**c**	48.	**a**	73.	**c**	98.	**e**
24.	**d**	49.	**b**	74.	**c**	99.	**c**
25.	**a**	50.	**d**	75.	**a**		

Level III Questions
Neutron Radiographic Testing Method

Physical Principles

Nature of Penetrating Radiation

1. Neutrons:

   a. are uncharged particles
   b. have a mass slightly less than a proton
   c. are found in the nuclei of all isotopes
   d. have all of the above properties

   B.13

2. Which of the following is true for neutron radiographic conditions?

   a. hydrogen in the object will more nearly stop high energy neutrons (well above 0.025 eV) than it will typical thermal neutrons (close to 0.025 eV)
   b. epithermal neutrons (of indium resonance energy) will cause more ionization per neutron than thermal neutrons and therefore cause more risk from radiation damage to sensitive materials
   c. in neutron radiography of a typical fast reactor fuel pin, each thermal neutron captured normally gives rise to about two fast neutrons and significant delayed gamma radiation
   d. the angle of scatter for fast neutrons will, on the average, be greater than that of thermal neutrons and this is one factor contributing to the difficulty of clear fast neutron radiography

   K.3

3. A thermal neutron beam is extracted horizontally from a reactor. Neglecting scatter by air or other materials in its path, what would you expect the beam path over the first 15 m (49 ft) to do?

   a. remain horizontal because the neutrons are essentially unaffected by gravity
   b. drop by 23 mm
   c. drop by 228 mm
   d. drop by 15 mm (0.6 in.)

   K.14-15

Interaction between Penetrating Radiation and Matter

4. When the mass absorption coefficient for neutrons is plotted as a function of atomic number of absorber, it presents a:

   a. regularly increasing curve
   b. random curve
   c. single peak curve
   d. double peak curve

   A.4

5. You have to neutron radiograph an object that required visualization of very small hydrogenous details behind 76 mm (3 in.) of steel. The standard technique using a tangential beam on a reactor proves insufficient. What would be the recommended approach?

   a. send the object to a high intensity (or flux boosted) Cf-252 system where it could be exposed for very long periods (e.g., over a weekend)
   b. send the object to a Van de Graaff neutron radiography system and request fast neutron point source neutron radiography using sulphur activation transfer to avoid any gamma fog
   c. send the object to a facility performing cold neutron radiography
   d. send the object to a pulsed reactor and use real time imaging in conjunction with time of flight methods

   S.16

6. Which of the following statements is incorrect?

   a. neutron radiography excels in imaging plastic objects behind metal walls
   b. X-radiography is eminently suitable for imaging complicated nonradioactive metal objects that are not too thick (i.e., less than 1 cm [0.4 in.])
   c. neutron radiography is suitable for imaging items composed of iron and plastic regardless of the thickness of the plastic provided only that the iron is less than 25 mm (1 in.) thick
   d. indium resonance neutrons are frequently used to enhance the penetration of thick plastic objects

   A; S

7. From the following penetrating radiations, which one can best be used to radiograph a cylinder of hydrocarbon plastic that is 50 mm (2 in.) in diameter in which we want to image embedded metal parts?

   a. fast neutrons
   b. thermal neutrons
   c. cold neutrons
   d. X-rays

A; S

Imaging by Film

8. Following neutron capture in a gadolinium-foil converter, which of the following is emitted from the foil surface interact in the photographic emulsion to produce a high-resolution latent image?

   a. X-rays
   b. gamma rays
   c. alpha particles
   d. conversion electrons

A.49-50

9. A fact concerning the reciprocity law failure in neutron radiography is that:

   a. film density of X-ray films is not proportional to log 10 exposure
   b. film density of light sensitive films is not proportional to exposure
   c. reciprocity law failure of X-ray films is partially alleviated by development at higher temperatures (80 °F)
   d. reciprocity law failure of light sensitive films is partially alleviated by exposure with scintillator convertors at low temperature

S.42

10. When a neutron-sensitive fluorescent screen is used in conjunction with a photographic film, the film exposure satisfies the Schwarzchild relationship, $E = I + p$, where $p = 0.75$. This means that, if a flux, I_0, gives a satisfactory exposure in 1 minute, a flux 1/10 as great, $I_0/10$, will require an exposure time of approximately:

    a. 5 minutes
    b. 10 minutes
    c. 20 minutes
    d. 40 minutes

F.349

11. The photographic latent image may be defined as that radiation-induced change in a grain or crystal that renders the grain readily susceptible to the chemical action of a developer. The material of the latent image itself is typically:

    a. silver ions
    b. silver atoms
    c. silver oxide
    d. silver iodide

D.141

12. Track etch neutron imaging differs from dysprosium transfer imaging in which of the following ways?

    a. it provides higher contrast
    b. it is unaffected by gamma fogging from the beam or object
    c. it provides poorer edge definition
    d. it is suitable for use in low intensity beams where long exposure times are acceptable

Y.584-586

Imaging by Fluorescent Materials

13. When fluorescent materials are employed for neutron radiographic imaging, reciprocity failure is often encountered. This is due to the fact that:

    a. greater energy is deposited in the film grain by light emission from the fluorescent screen to render the grain developable than can be achieved with metal foil conversion screens
    b. less energy is deposited in the film grain to render it developable than with metallic conversion screens
    c. the fluorescent screen is electrically nonconducting
    d. the fluorescent screen does not respond to gamma radiation

Imaging by Electronic Devices

14. Electron imaging devices invariably provide:

    a. amplification of both signal and noise including statistical noise
    b. intimate film to conversion screen contact
    c. discrimination against scattered neutrons
    d. direct external conversion electrons of varying energies

A.72-76

15. Substitution of gadolinium-oxysulfide for ZrS-Li^6F as the input converter is an electronic imaging device will normally result in:

    a. increased discrimination against gamma radiation
    b. increased inherent unsharpness
    c. the need for increased electronic gain
    d. decreased image retention between frames

X

Radiometry

16. The "cadmium ratio" of a neutron refers to:

    a. the ratio of thermal neutron flux above cadmium cutoff
    b. the ratio of thermal neutron flux to gamma dose rate
    c. the ratio of total neutron flux to the neutron flux above about 0.5 eV
    d. the ratio of neutron flux above about 0.5 eV to the thermal neutron flux

A.33

SOURCES OF NEUTRONS
Neutron Sources—General Reactors

17. High quality neutron radiographs are produced by reactor neutron sources because of the:

    a. absence of fast neutrons
    b. large physical size of the reactor core
    c. strong resonance in the reaction cross section at 120 keV
    d. high core flux in the range of 10^{12} to 5×10^{14} n/cm^2-s

A.24; CC.175

Accelerators

18. Which one of the following generators has a significant neutron yield at low bombardment energies?

    a. Van de Graaf
    b. T(d,n)
    c. linear accelerator
    d. D(d,n)

CC.178

Isotopic Sources

19. The neutron-producing reaction of Cf-252 is:

    a. spontaneous fission
    b. (α, n) reaction
    c. (γ, n) reaction
    d. none of the above

A.24

20. Of the isotopic sources employed for neutron radiography, which of the following sources decays by the spontaneous fission process?

    a. Sb-124-Be
    b. Po-210-Be
    c. Cf-252
    d. none of the above

A.24

RADIATION DETECTORS
Imaging Converter Screens

21. For fast films, thermal neutron fluxes of about 10^4 nV (n/cm^2-s) represent the minimum useful intensity for conventional transfer neutron radiography utilizing which one of the following conversion screens?

    a. rhodium
    b. dysprosium
    c. silver
    d. gadolinium

S.38-39

22. You normally use gadolinium direct exposure methods at your facility. You have available only one gadolinium screen, 36 cm × 43 cm (14 × 17 in.). One day, you wish to neutron radiograph a single very large object (100 cm × 30 cm [39 × 12 in.]) in a single exposure. Obtaining a widely divergent beam is no problem at this facility. Screens, however, are a problem. You are willing to sacrifice some image quality in order to avoid the expense of such a large vapor deposited screen. To get the best results at the lowest cost you would:

    a. use a thick plastic screen as a proton knock-on neutron converter
    b. use an aluminum plate coated with gadolinium oxide paint
    c. use "no-screen" type film, with no converter but with approximately treble normal neutron exposures
    d. use an europium screen

S.35-55

Film-principles, Properties, Use with Neutron Converter Screens

23. A 20 percent increase in exposure ($\Delta \log E = 0.08$) for a given film-convector combination results in a film density change from 2.3 to 2.7. The gradient of the characteristic curve is:

   a. 0.2
   b. 0.032
   c. 2.0
   d. 6.0

D.125

24. LiF, Zn S(Ag) neutron scintillators, which can be used as a fluoroscopic screen, emit blue light that:

   a. is a good match for the spectral response of most film
   b. is a good match for the spectral sensitivity of the human eye
   c. creates high screen brightness
   d. creates high resolution in the image

F.72

25. Which of the following facts are true concerning the use of a regular X-ray gadolinium$_2$0$_2$S screen as a neutron converter?

   a. the film is exposed by a combination of light and electrons generated initially at over 10 different energy levels
   b. the film is exposed mainly by blue light as for a typical LiF-ZnS (Ag) scintillator screen
   c. the gamma discrimination is better than would be true for a gadolinium metal foil converter
   d. the sharpness of the radiograph will usually be inferior if the rear scintillator face is photographed through a camera rather than imaged with an equal quality film placed adjacent to the rear of the screen

A.72

Fluoroscopy—TV and Optical Systems

26. A practical video neutron-radiographic system could not be based on a:

   a. boron converter
   b. dysprosium converter
   c. gadolinium converter
   d. lithium converter

F.348

Gaseous Ionization Detectors

27. When using a gaseous proportional counter for neutron radiography, the principal ionization event utilizing thermal neutrons is due to which of the following?

   a. photoelectric effect with certain gases
   b. piezoelectric effect on the electrodes
   c. gaseous discharge between the electrodes
   d. alpha particles

S.46

Neutron Detectors

28. Which of the following types of detectors would be least likely to be used for neutron detection?

   a. BF_3 proportional counter
   b. NaI scintillation counter
   c. plastic scintillation counter
   d. lithium-glass scintillation counter

F

29. Neutrons may be detected and measured using a foil activation technique. A foil material suitable for detecting thermal neutrons with this technique is:

   a. carbon
   b. gold
   c. gadolinium
   d. boron

30. A boron trifluoride neutron counter is known as a proportional counter because:

   a. the neutron cross section is proportional to the inverse of neutron energy
   b. the neutron cross section is proportional to the inverse of neutron velocity
   c. the electronic signal is proportional to the voltage applied to the system
   d. the quantity of charge collected for a given amount of radiation is larger than, but proportional to, the original amount of ionization

K.279

Instrumentation

31. The output of most non-imaging neutron detectors is:

   a. an electric signal
   b. an acoustic signal
   c. a mechanical signal
   d. a thermal signal

O.318

Gaging and Control Processes

32. A gaging setup can distinguish 5% changes in intensity of a transmitted neutron beam. If the average thickness of a sheet being gaged attenuates the neutron-beam intensity by a factor of 20 (~ 3 mean free paths), how small a change in thickness will be detectable?

    a. 0.4%
    b. 1.15%
    c. 1.7%
    d. 4%

S; O.318

33. A system of gaging in which the radiation is collimated and projected through a test item and the quantity of unabsorbed radiation is measured is referred to as the:

    a. fluoresence method
    b. absorption differential method
    c. attenuation "build up" method
    d. transmission method

S.58

34. Generally, the sensitivity and accuracy of thickness gaging by reflection methods is:

    a. superior to transmission gaging
    b. superior to fluorescence methods
    c. inferior to transmission gaging
    d. approximately the same as with transmission gaging

S.59

PERSONNEL SAFETY AND RADIATION PROTECTION

Exposure Hazards

35. Where isotopic sources are employed for field neutron radiographic applications, the radiation levels near the biological shield should not exceed:

    a. 200 mR/h at the surface and 2 mR/h at 1 m (3.3 ft)
    b. 700 mR/h at the surface and 10 mR/h at 1 m (3.3 ft)
    c. 1 R/h at the surface and 20 mR/h at 1 m (3.3 ft)
    d. none of the above

S.97

Specific Neutron Hazards

36. Which of the following neutron personnel monitoring conditions is incorrect?

    a. monitoring of thermal neutrons (< 0.5 eV) does not pose a serious problem because of the ready availability of a number of suitable detectors, such as the cadmium- or rhodium-covered personnel-monitoring films and LiF-6 thermoluminescent detectors
    b. the detector should respond predominantly to the type of radiation which it is meant to monitor and its sensitivity to other types of radiation that may be simultaneously present should be minimal
    c. intermediate energy neutron monitoring does not pose a serious problem because the many types of albedo dosimeters now in use are reasonably reliable due to the fairly linear response independent of neutron spectrum

T.343-345

Operation and Emergency Procedures

37. "Occupational dose" as applied to the exposure of an individual to radiation in a restricted area or in the course of employment in which the individual's duties involve exposure to radiation shall:

    a. not be deemed to include any exposure of an individual to radiation for the purpose of medical diagnosis or medical therapy of such an individual
    b. include all radiation received by an individual from his work duties, medical diagnosis, dental, or medical therapy
    c. include only that radiation which he deems necessary to report
    d. include only those doses that exceed 5 R in any one year

J

38. The rad, as used in USNRC Title 10, Chapter 1 CFR, Part 20, is a measure of dose of any ionizing radiation to body tissue in terms of the energy absorbed per unit mass of the tissue. One rad is the dose corresponding to the absorption of how many joules (ergs) per gram of tissue?

    a. 8.3 μJ (83.6 ergs) of energy
    b. 10 μJ (100 ergs) of energy
    c. 9.3 μJ (93 ergs) of energy
    d. 7.6 μJ (76.2 ergs) of energy

J

39. An unrestricted area means:

a. any area where doctors and medical staff personnel are allowed to pass while on duty
b. any area established to perform radiographic operation usually roped off to 2 mR/h
c. any area used to store radioactive materials or radiation-producing equipment
d. any area to which access is not controlled by the licensee for purposes of protection of individuals from exposure to radiation and radioactive materials, and any area used for residential quarters

J

40. Units of radioactivity are commonly measured in terms of disintegrations per unit time or in curies. One curie is:

a. 2.2×10^3 disintegrations per second
b. 3.7×10^6 disintegrations per second
c. 3.7×10^{10} disintegrations per second
d. dependent on the type of isotope

41. What is the maximum dose that a worker can be permitted to receive in a restricted area in any one calendar quarter?

a. 5 rems — whole body, head trunk, active blood-forming organs, lens of the eye, or gonads
b. 4 rads — whole body, head trunk, active blood-forming organs, lens of the eye, or gonads
c. 3 rems — whole body, head trunk, active blood-forming organs, lens of the eye, or gonads
d. 1-1/4 rems — whole body, head trunk, active blood-forming organs, lens of the eye, or gonads

THE RADIOGRAPHIC PROCESS

Imaging Considerations - Sensitivity

42. According to the standard ASTM E-545, which of the following is an acceptable level of neutron radiographic sensitivity?

a. S = 10
b. S = 15
c. R = 5
d. R = 10

C.3

Contrast and Definition

43. Which of the following parameters do not directly affect radiographic definition?

a. focal spot or source size
b. density
c. type of screen
d. radiation quality

D.66

44. Film contrast refers to:

a. the density difference in two adjacent regions of film
b. the steepness (slope) of the characteristic curve
c. the ratio of X-ray or gamma-ray intensities transmitted by two selected portions of a specimen
d. minimum perceptible density change

V.20-35

45. In general, the contrast of radiographic films (except those designed for use with fluorescent screens):

a. increases continuously with film density
b. decreases as the density is increased
c. remains practically unchanged for different density levels
d. is inversely proportional to film density

D.20-35

46. Specimens with uniform thickness and composition by definition have:

a. high subject contrast
b. good definition
c. high film contrast
d. low subject contrast

U.92

47. The sharpness of outline in a radiographic image is referred to as:

a. definition
b. sensitivity
c. latitude
d. contrast

D.67

48. Poor radiographic definition could be the result of:

a. focal spot size
b. source film distance
c. poor film-screen contact
d. all of the above

D.66

49. Based on the characteristic curves of the films below, which film provides the highest contrast?

    a. Film X
    b. Film Y
    c. Film Z
    d. cannot be determined from the curves

D.124

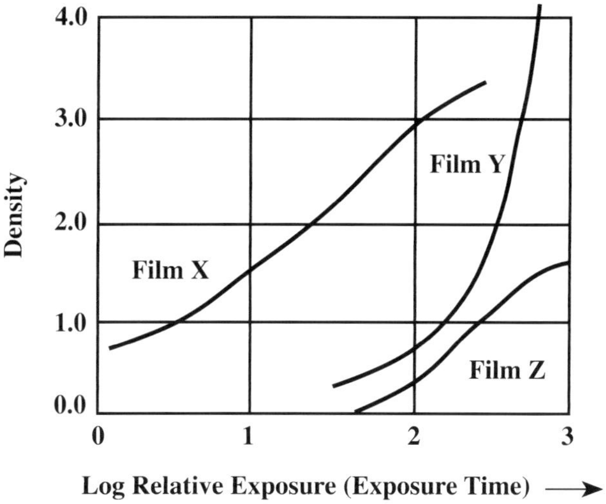

Geometric Factors

50. When a given resolution is desired in a radiograph, a compromise must be made between attainable resolution and neutron flux at the image plane because:

    a. a greater flux is available at the image plane in high-resolution work
    b. the flux available at the image plane is directly proportional to the solid angle subtended by the aperture at that plane
    c. the flux at the image plane is less
    d. the flux available at the image plane is inversely proportional to the solid angle subtended by the aperture at that plane

Z.81

51. Converter or intensifying screens are utilized in both the direct and indirect photographic techniques. Which of the following is arranged from the highest to the lowest cross section for thermal neutrons?

    a. gadolinium, dysprosium, indium, cadmium
    b. samarium, cadmium, dysprosium, gold
    c. dysprosium, indium, gadolinium, silver
    d. boron, samarium, cadmium, lithium

A.38

Intensifying Screens

52. Dy-164 has a cross section of 2 700 10^{-28} m^2 (2 700 barns) for the 1.25 minute half-life activity, and 800 10^{-28} m^2 (800 barns) for the 140 minute half-life activity. Considering activation transfer neutron radiography in which there is minimal delay between foil neutron exposure and film contact, and assuming a long transfer period (12 hours), which of the following is true?

    a. a greater film density could be obtained from a 20 MW-sec reactor pulse than from the same facility operated at 100 kW for 10^4 seconds (2.78 hours)
    b. the film density obtained from a 2 000 MW reactor pulse is approximately equal to that obtained from an operation of the same facility at 100 kW for 10^3 seconds (16.6 minutes)
    c. if now a cadmium foil filter is put in the incident beam (epithermal neutron radiography instead of thermal) and the long 10^4 second exposure is used first with dysprosium foil transfer and then with indium foil transfer, the indium transferred activity is greater for dysprosium
    d. none of the above

A.64-70

Scattered Radiation

53. The problem of scatter in an hydrogenous object being radiographed can be reduced by:

    a. increasing the L/D ratio
    b. using the transfer technique
    c. employing laminography
    d. placing a neutron-absorbing collimating grid between the object and the film

A.103

Source Factors

54. Given a neutron beam with an L/D of 250, a thermal flux of 10^6 n/cm^2-s and a 36-43 cm (14-17 in.) exposure area, what is the geometric sharpness for a test object of 6 mm (0.25 in.) thick?

    a. 0.01
    b. 0.05
    c. 0.001
    d. 0.005

R.60

55. Indium-resonance neutrons:

    a. are not used in conjunction with an indium converter
    b. are never used in transfer radiography
    c. are not used for reactor-fuel radiography
    d. none of the above

A.64-70

Detection Media

56. Which of the following converters cannot be used for transfer neutron radiography?

    a. indium
    b. dysprosium
    c. cadmium
    d. rhodium

57. Cold neutrons can be detected with high efficiency, for imaging by:

    a. ordinary gadolinium converters
    b. radiography film directly
    c. special beryllium filters
    d. both a and b

S.15-16

58. The relative n/γ-ray sensitivity of metal-screen converters in generally about __________ that of scintillating-screen converters for direct exposure radiography.

    a. two orders of magnitude greater than
    b. one order of magnitude greater than
    c. the same as
    d. one order of magnitude less than

F.350

Exposure Curves

59. The contrast observed in a radiographic image is greater when:

    a. the film density gradient is maximum
    b. the response of the human eye is logarithmic
    c. one operates at the lowest portion of the D-Log E curve
    d. one operates at the highest portion of the D-Log E curve

D.69

Film Processing

Darkroom Equipment and Chemicals

60. Which statement concerning darkroom equipment and chemicals is incorrect?

    a. processing rooms should be supplied with filtered air, at a pressure above that of the outside
    b. when exposed film is placed in developer, the solution penetrates the emulsion and begins to transform the exposed silver halide crystals to metallic silver
    c. vessels used for processing of films should not be made from aluminum or tin because they could cause contamination and result in fog in the radiograph
    d. the usefulness of a fixer solution is ended when it has lost its basicity

D.80-96

61. In automatic processing, over-replenishment of the fixer solution may result in:

    a. poor fixation
    b. insufficient hardening
    c. failure of the film to be transported through the fixer rack
    d. none of the above

D.90

Background Lighting

62. The contrast sensitivity of the human eye is greatest when the surroundings compared to the area of interest on a radiograph are:

    a. about the same brightness
    b. of a lower brightness
    c. of a higher brightness
    d. brightness is not a factor in contrast sensitivity

V.17-21

63. For best contrast sensitivity, the film viewing room should have lighting:

    a. as dark as possible
    b. of approximately 38 lumens
    c. of approximately 70 lumens
    d. as light as the area of interest in the film being reviewed

D.73

64. When reviewing film, background lighting should:

    a. be virtually eliminated
    b. not reflect on the film under examinations
    c. be carefully filtered
    d. be approximately 20 lumens

D.73

Density—Judging Radiographic Quality

65. Photographic density refers to the quantitative measure of film blackening. When no danger of confusion exists, photographic density is usually spoken of merely as "density." Density is defined by the equation:

    a. $D = \frac{I_t}{I_0}$
    b. $D = \log \frac{I_0}{I_t}$
    c. $D = \frac{I_0}{I_t}$
    d. $D = \log \frac{I_t}{I_0}$

    where:
    D is density,
    I_t is the light intensity transmitted and,
    I_0 is the light intensity incident on the film

Contrast

66. Radiographic contrast for neutron radiography is a function of the:

    a. size and shape of the neutron source
    b. length of the beam collimator
    c. L/D ratio
    d. gradient of the characteristic curve of the film and the neutron energy

D.30

67. Radiographic contrast in neutron radiography can sometimes be increased by proper choice of converter foils. If a gadolinium foil were chosen in place of an indium foil, you might be seeking contrast enhancement from:

    a. the increased cross section in the sample for lower energy neutrons
    b. the use of prompt rather than delayed emission of radiation
    c. the generally thinner gadolinium foil
    d. all of the above

A.49, 69

Definition

68. Standard ASTM E-545 uses a maximum of how many sensitivity indicators?

    a. 2
    b. 3
    c. 4
    d. 5

C.3

Artifacts

69. The appearance of colored stains on a processed radiograph could be caused by:

    a. neutralization of the acid in the fixer solution
    b. neutralization of the alkaline content in the fixer solution
    c. acidification of the fixer solution
    d. underdevelopment

D.86

70. Prolonged washing of film in water above 68 °F has a tendency to:

    a. crystallize the gelatin
    b. soften the gelatin
    c. cause the yellow stain
    d. cause the image to fade

D.86

71. The ASTM beam purity indicator is designed primarily to indicate:

    a. contrast
    b. sharpness
    c. speed
    d. all of the above

C.7

72. The image quality indicator is generally made of:

    a. the same material as that of the specimen being radiographed
    b. a more dense material than that of the specimen being radiographed
    c. a material having a 10% greater density than the specimen being radiographed into account thickness changes
    d. none of the above

C

Exposure Calculations

73. You design a divergent collimator NR system on a given beam port such that the exposure area is 10 × 10 cm (4 × 4 in.), and the geometric unsharpness is 50 μ at 1 cm (0.4 in.) separation between detail and screen. Using the same neutron source beam tube, you now redesign for a 30 × 30 cm (12 × 12 in.) exposure area and a geometric unsharpness of 100 μ at 1 cm (0.4 in.). The larger aperture and divergence cause a flux depression at the source of a factor of 2 as compared with the first system. The new exposure time required when using gadolinium foil and Kodak SR film on a thin object is 300 seconds. Estimate the original source flux and exposure time.

    a. 2×10^{11} n/cm^2-sec, 6 000 seconds
    b. 2×10^{12} n/cm^2-sec, 600 seconds
    c. 2×10^{13} n/cm^2-sec, 300 seconds
    d. 2×10^{14} n/cm^2-sec, 30 seconds

AA

74. In order to provide the initial evaluation as to the practicality of a particular direct neutron radiograph exposure method, which term is not needed?

    a. number of nuclei/cm^3 of absorbing material
    b. decay constant (0.693/$T_{1/2}$ where $T_{1/2}$ is the half-life)
    c. total cross section of material (in cm)
    d. material thickness (in cm)

A.109

RADIOGRAPHIC TECHNIQUES

Blocking and Filtering

75. A filter material commonly used in a neutron beam to improve the neutron-to-gamma ratio is:

    a. U-238
    b. bismuth
    c. copper
    d. lithium

A.82

Multifilm Techniques

76. Which of the following is the most inclusive?

    a. M film
    b. single emulsion film
    c. Class I film
    d. AA film

V.26ff

Enlargement and Projection

77. With the normal divergent collimator, image magnification occurs as the spacing between the object plane and the detector plane increases. If *L* is the source-to-object distance, and *t* the object-to-image plane distance, then the magnification is:

    a. L/t
    b. t/L
    c. $L/L-t$
    d. $(L+t)/L$

DD.5-8

Autoradiography

78. A practical application of autoradiography subsequent to neutron activation is:

    a. a study of hydrogen embrittlement of titanium
    b. faded photograph restoration
    c. a study of the diffusion of boric acid in concrete samples
    d. both b and c

B.512

Flash Radiography

79. High-speed radiography can be best performed by:

    a. autoradiography
    b. in-motion radiography
    c. laminography
    d. flash radiography using a pulsed neutron source

D.12

Fluoroscopy

80. One of the main disadvantages of a fluoroscopic system is:

    a. extensive operator training required
    b. high image brightness on screen
    c. flaw image enlargement
    d. limited ability to see fine detail

W.149

81. Fluoroscopy possesses certain advantages over film radiography such as its:

    a. attainable resolution is greater than with film
    b. sensitivity is far greater than can be obtained with film
    c. low cost and high speed
    d. none of the above

S.42

Microradiography

82. Which of the following parameters has minimal or no effect on resolution when making a microradiograph?

   a. the angular divergence of the neutron beam
   b. the thickness of the object being investigated
   c. the type of converter material and the converter screen thickness
   d. the latitude of the photographic emulsion

BB.139

Laminography (Tomography)

83. In tomography, the plane of the object that will be in focus is determined by the:

   a. distance of the source from the object
   b. ratio of the object-film distance to the object-source distance
   c. position of the pivot point of the film-source mechanism relative to the object
   d. angle through which the film-source mechanism travels

D.103

Gaging

84. Isotopic sources have certain advantages over radiation generators for gaging in that:

   a. an isotopic source yield is stable over the short term, although it does decay with time
   b. these sources require more stringent regulation than accelerators to protect personnel
   c. isotopic sources can be switched off with ease
   d. none of these

A.20

Real-Time Imaging

85. Which of the following is not associated with real-time neutron imaging?

   a. optical coupling
   b. television camera
   c. intensifier tube
   d. dysprosium foil

S.43-44

Image Analysis Techniques

86. Assuming 10^6 electrons associated with a particular feature of an image, the expected variation due to statistical consideration is:

   a. 100 electrons
   b. 500 electrons
   c. 1 000 electrons
   d. 5 000 electrons

A.74

RADIOGRAPHIC INTERPRETATION

Image-Object Relationships

87. The best standard for image-object relationship is:

   a. a comparison standard
   b. a BPI standard
   c. a penetrameter
   d. a type A resolution standard

C.paragraph 3.2

Material Processing as It Affects the Use of Item and Test Results Material Considerations

88. Radiographic detection of small amounts of explosive in thick steel will give best contrast if what kind of neutrons are used?

   a. cold
   b. thermal
   c. epithermal
   d. the optimum resonance energy
   e. fast

F.356; S.50

89. Compared to thermal neutrons, in crystalline materials, cold neutrons are:

   a. scattered more and absorbed less
   b. scattered less and absorbed more
   c. scattered less and absorbed less
   d. scattered more and absorbed more

F.356

Discontinuities: Their Causes and Effects

90. Particles of foreign material such as sand or slag that are embedded in the cast metal are indicative of:

    a. gas holes
    b. buckles
    c. rattails
    d. inclusions

W.309

91. One lower limit of detectability of microporosity in materials can be caused by:

    a. an L/D ratio that is too small and/or clumping of the grains of the developed film
    b. variation in the neutron source flux at the image plane
    c. an excessively high L/D ratio
    d. none of the above

A.42, 95ff

Codes, Standards, Specifications, and Procedures

92. As stated in the "Standard Method for Determining Image Quality in Thermal Radiographic Testing," ASTM E-545-75, which of the following statements are justification for preparing a process control radiograph for verification of specified exposure and sensitivity requirements?
    (1) The part setup does not provide for location of image quality indicators to give adequate readings.
    (2) The part configuration necessitates a film-to-beam orientation which does not provide satisfactory density measurements for calculation of exposure by collimated thermal neutrons.
    (3) The size or setup of parts is such that the background film density variation across the film exceeds 5% (±2.5% variation from the numerical mean of 5 measurements).
    (4) The part materials necessitate the use of a conversion screen that responds to neutrons of energies other than thermal in order to give adequate readings.
    (5) The size or setup of parts is such that the object-scattered neutron level relative to background density, sensitivity, or facility-scattered neutron exposure requirements is cause for nonconformance.

    a. (1), (2), and/or (3)
    b. (2), (3), and/or (5)
    c. (1), (2), and/or (4)
    d. (1), (2), and/or (5)

C.2-4

Level III Answers
Neutron Radiographic Testing Method

1. **a**
2. **c**
3. **c**
4. **b**
5. **c**
6. **c**
7. **d**
8. **d**
9. **d**
10. **c**
11. **b**
12. **d**
13. **b**
14. **a**
15. **c**
16. **c**
17. **d**
18. **b**
19. **a**
20. **c**
21. **b**
22. **b**
23. **d**
24. **a**
25. **a**
26. **b**
27. **d**
28. **b**
29. **b**
30. **d**
31. **a**
32. **c**
33. **d**
34. **c**
35. **a**
36. **c**
37. **a**
38. **b**
39. **d**
40. **c**
41. **c**
42. **d**
43. **b**
44. **b**
45. **a**
46. **d**
47. **a**
48. **d**
49. **b**
50. **b**
51. **b**
52. **c**
53. **d**
54. **c**
55. **d**
56. **c**
57. **a**
58. **d**
59. **a**
60. **d**
61. **d**
62. **a**
63. **d**
64. **b**
65. **b**
66. **d**
67. **a**
68. **c**
69. **a**
70. **b**
71. **a**
72. **a**
73. **b**
74. **b**
75. **b**
76. **c**
77. **d**
78. **b**
79. **d**
80. **d**
81. **c**
82. **d**
83. **c**
84. **a**
85. **d**
86. **c**
87. **a**
88. **a**
89. **b**
90. **d**
91. **a**
92. **d**